Linda GOUISSEM

Methods for incorporating $CaCO_3$ into HDPE composites

Linda GOUISSEM

Methods for incorporating $CaCO_3$ into HDPE composites

Impact of direct and concentrate methods on mechanical, thermal and processing properties

ScienciaScripts

Imprint
Any brand names and product names mentioned in this book are subject to trademark, brand or patent protection and are trademarks or registered trademarks of their respective holders. The use of brand names, product names, common names, trade names, product descriptions etc. even without a particular marking in this work is in no way to be construed to mean that such names may be regarded as unrestricted in respect of trademark and brand protection legislation and could thus be used by anyone.

Cover image: www.ingimage.com

This book is a translation from the original published under ISBN 978-620-6-72926-6.

Publisher:
Sciencia Scripts
is a trademark of
Dodo Books Indian Ocean Ltd. and OmniScriptum S.R.L publishing group

120 High Road, East Finchley, London, N2 9ED, United Kingdom
Str. Armeneasca 28/1, office 1, Chisinau MD-2012, Republic of Moldova, Europe
Managing Directors: Ieva Konstantinova, Victoria Ursu
info@omniscriptum.com

Printed at: see last page
ISBN: 978-620-8-37100-5

Thanks

Above all, I thank God, the Almighty, for giving me the strength and opportunity to complete this work.

I'd like to express my gratitude to my family, who have been with me every step of the way. I would especially like to thank my husband, Abdelhakim, for his invaluable help, patience and commitment.

I would like to express my gratitude to the heads of the Emerging Materials Research Unit and the Institut d'Optique et de Mécanique de Précision laboratories for giving me the opportunity to carry out the characterization tests essential to this work.

Contents

General introduction

General introduction

It's hard to believe just how important plastics have become in our daily lives. It seems we've always known about these materials, and we tend to take it for granted that they're present all around us every day, for example in our clothes, the pen we write with, the chair we sit on or the packaging on the food we eat. Let's take a look around us: how much plastic do we use? So it's sometimes hard to believe that plastics have only been commonly available for about a century. Yet in that time, the impact they have had on our quality of life and the products we have access to has been enormous. Plastics give us the ability to make beautiful, well-designed products from the many different types of plastic available today. In the field of manufacturing technologies, there is a very high degree of technological mastery of plastics and a series of sophisticated technological processes that enable us to make and shape them in many different ways. [1,2]

Plastics are initially produced as powders, granules or laces (like straws). However, they are rarely used in their pure state, but are modified by the addition of other materials that give them their particular properties.

In this context, the development of mineral-filled polymer composites has advanced significantly in recent decades; these materials, in which a polymer matrix is reinforced by mineral fillers, are distinguished for several reasons. Their enhanced mechanical properties make them particularly useful in demanding applications such as automotive, aerospace and consumer durables. In addition, their thermal properties increase heat resistance, helping composites to retain their mechanical properties at higher temperatures. In economic terms, the use of mineral fillers often reduces the overall cost of composite materials, enabling production costs to be cut while maintaining, or even improving, the performance of the end product. **[3]** These composites have a wide range of applications, from electronic components and packaging to the construction industry and sports equipment. Their ability to be molded into complex shapes with an excellent surface finish makes them highly versatile. [4]

Ongoing research in this field explores new combinations of polymers and mineral fillers, as well as advanced processing techniques to further enhance composite performance. These advances make mineral-filled polymer composites particularly attractive for applications requiring a good compromise between performance and cost, while paving the way for new innovations in materials. [4]

Among these composites, those based on high-density polyethylene (HDPE) and calcium carbonate ($CaCO_3$) stand out for their many applications in sectors such as packaging, construction and automotive. [5] HDPE is renowned for its strength, chemical resistance and ease of processing, while $CaCO_3$, as a mineral filler, offers advantages such as improved rigidity and reduced costs. [6]

Traditionally, HDPE/$CaCO_3$ composites are prepared by a direct method where $CaCO_3$ is mixed directly with the HDPE. However, this approach can lead to uneven dispersion of $CaCO_3$ particles in the polymer matrix, which can adversely affect the final properties of the composite. [7] To remedy this problem, one method that is well established in the industry is to prepare a $CaCO_3$ concentrate (Masterbatch) before incorporating it into HDPE. This technique allows better control of the dispersion of $CaCO_3$ particles, ensuring a more homogeneous distribution within the polymer matrix.

In this study, we prepared the $CaCO_3$ (CC) concentrate using an internal mixer, specifically a plastograph, to ensure efficient initial dispersion of the particles in the HDPE matrix. The HDPE/CC composites were also elaborated in an internal mixer, promoting homogeneous mixing and optimal material processing.

The main objective of our research is to assess the impact of incorporating $CaCO_3$ from concentrates on the properties of HDPE/$CaCO_3$ composites. By comparing this method with conventional preparation, we aim to highlight potential advantages in terms of mechanical, thermal and processing properties. We also wish to demonstrate to students and researchers that this method is not only effective, but worthy of adoption in their future research. The results of this study could thus offer promising prospects for the optimization of HDPE- and $CaCO_3$-based composites, while contributing to a reduction in production costs.

This work is divided into three chapters:

The first chapter presents a review of the literature on plastics, with particular emphasis on polyethylene and reinforcing fillers such as calcium carbonate, as well as a section dedicated to techniques for preparing polymer/filler mixtures. The second chapter describes the materials used, analysis methods and characterization test conditions, presented in detail. The third chapter is devoted to analysis and discussion of the results. Finally, this work concludes with a general summary and a discussion of the prospects opened up by this study.

References

[1] : Chemical Safety Facts. "*Introduction to Plastics*" [online]. Available from: <http:// plastics.americanchemistry.com/Introduction-to-Plastics> (accessed: 7/04/2022).

[2] : N. J. MILLS. 2005, "Plastics." 3rd Edition Ch 01, England, Final Proof, 413 p.

[3] : M. SANGRONIZ et al. "A Review on the mechanical and thermal properties of polymer nanocomposites. A Review on the mechanical and thermal properties of polymer nanocomposites." Polymers, Vol. 10, 2019, pp. 979-991.

[4] : F. P. LA MANTIA and M. MORREALE. "Economic and environmental benefits of polymer Matrix composites in automotive applications." Composite Structures, Vol. 92, 2010, pp. 1153-1159.

[5] : H. KIM, H. PARK and D. LEE. "Advances in polymer composites filled with mineral particles: mechanical, thermal, and economic properties." Journal of Composite Materials, Vol. 52, 2018, pp. 2007-2021.

[6] : Abb. chemical resistance guide. [in press]. Available at: <https://library.e.abb.com >. (Accessed: 10/07/2024).

[7] : R. ROTHON and C. PAYNTER. "Fillers for polymer applications." Polymers and Polymer Composites, Vol. 35, 2017, pp. 1203-1211.

Chapter I: Bibliography

Introduction

A polymer is a macromolecule made up of a large number of repeating units called monomers, linked together by covalent bonds. Depending on their origin, polymers can be natural (cellulose, proteins, DNA), derived from natural polymers by chemical modification, or synthetic, obtained by polymerization of monomers. [1]

Synthetic polymers, commonly referred to as plastics, underwent major development in the 20th century, replacing many traditional materials such as wood and metal. Thanks to their varied properties (lightness, durability, ease of shaping, low cost), they have become ubiquitous in everyday life, and find numerous industrial applications. [2]

Polymers are classified into three main categories according to their thermomechanical properties: [3]

- Elastomers (rubber) ;
- Thermoplastic polymers (polyethylene, polystyrene, polyvinyl chloride);
- Thermosetting polymers (epoxy resins, phenoplastics).
- Their properties are influenced by the nature of the monomers, their linkage and the structure of the polymer (linear, branched, cross-linked). Current research is focusing on the development of new polymers with tailor-made properties, to meet the growing needs of consumers. [2]

I.1 Polyethylene

I.1.1 Polyethylene in general

Polyethylene (PE) is a widely used polymer thanks to its remarkable mechanical, chemical and electrical properties. Its semi-crystalline molecular structure results from the alignment of segments (CH_2-) along the polymer chain. This regular organization favors the formation of crystalline zones, giving PE high strength and rigidity. [4]

Polymer chain length is a crucial factor in determining PE properties. In general, long chains favor the formation of extended crystalline structures, thus strengthening the material. Conversely, shorter chains lead to reduced crystallinity, conferring greater flexibility and making the material softer and less rigid. [4]

Thus, by adjusting the length of polymer chains during the manufacture of polyethylene, it is possible to modulate its properties to meet specific requirements, whether these involve high mechanical strength, increased flexibility, or a combination of these characteristics. [4]

The chemical structure of PE is shown in **Figure I.1** :

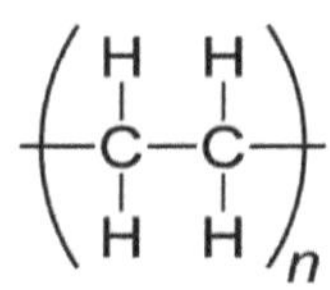

Figure I.1: Polyethylene repeat pattern [5].

I.1.2 High-density polyethylene synthesis

High-density polyethylene is mainly obtained by catalytic polymerization of ethylene at low pressure, using two types of catalyst: [6]

- Ziegler-Natta catalysts, developed in the 1950s by Karl Ziegler. They enable the synthesis of linear HDPE [7, 8].
- Chromium catalysts (Phillips process), developed in the 1950s by Phillips Petroleum. They produce HDPE with an average molecular weight distribution. [7, 9]

Two main polymerization processes are used: [6]

- Suspension or solution polymerization, in the presence of a solvent. This process requires removal of the solvent after polymerization.
- Fluidized-bed gas-phase polymerization. This more recent process makes it possible to obtain higher-capacity reactors (up to 200 kt/year) without using solvents.

The resulting HDPE has a linear, low-branched structure, favoring a high degree of crystallinity. This gives it great rigidity and mechanical strength, unlike low-density polyethylenes, which are more branched. [6]

I.1.3 Main properties of HDPE

High-density polyethylene is a thermoplastic polymer widely used for its excellent mechanical and chemical properties. It has high tensile, flexural and impact strength, making it suitable for a variety of applications requiring structural strength. In addition, HDPE offers good resistance to chemical corrosion, various chemicals and moisture, making it the material of choice for harsh environments. It is also valued for its good electrical insulation properties and ease of processing through processes such as extrusion, injection molding and blow molding. [10]

In terms of worldwide production, the most recent data indicate that several tens of millions of tonnes of HDPE are produced every year. [10]

Production continues to grow due to sustained demand in various industrial sectors, including packaging, construction, automotive and consumer goods. [10]

Table I.1: Main properties of HDPE [11]

Properties	Description
Security	No explosion in the event of fire ;
Mass gain	30-50% weight savings compared to other materials;
Freedom of form	Great freedom of shape to make the best use of the available volume;
Shock resistance	Good impact resistance even at low temperatures;
No corrosion	HDPE does not corrode;
Fuel permeability	Main weakness, requiring waterproofing techniques ;
Waterproofing techniques	- Fluorination of the inner surface of tanks ; - Sulfonation; - Coextrusion with other materials (polyamide, EVOH); - Blends with other polymers.

I.1.4 Use of HDPE in composites

High-density polyethylene is used in the manufacture of composites for a variety of applications: [12]

a) Coextrusion with other materials

HDPE can be co-extruded with polymers such as polyamide or EVOH (ethylene vinyl alcohol copolymer) to waterproof HDPE tanks. Although difficult and costly, this technique is effective in reducing fuel permeability, which is HDPE's main drawback. [12]

b) Blends with other polymers

The use of HDPE and polyamide blends is another simple, albeit expensive and moderately effective, method of improving impermeability. [12]

c) Thermoplastic matrix composites

Thermoplastic matrix composites, such as HDPE, can be used to produce more complex parts using direct manufacturing processes such as impregnation. Their mechanical behavior is studied and compared with that of thermoset matrix composites. [13]

d) Reinforced hybrid composites

HDPE can be used in the design of fiber-reinforced hybrid composites to achieve high mechanical performance. The fibers carry most of the stress, while the HDPE matrix serves mainly as a support. [14]

I.2 Expenses

I.2.1 Definition

A filler is any compound in powder, fiber or film form that is immiscible with the polymer. When added to the polymer, it improves the mechanical and thermomechanical properties of the final material into which it is incorporated. [15]

These fillers can be solid, immiscible substances dispersed by mechanical means in a matrix. They are introduced into a mixture to, among other things, reduce cost, improve properties or modify density. [16]

I.2.2 Load structure

The structure of a charge is defined on different scales. On a small scale, primary particles appear as assemblies of indivisible elements. These particles can then associate to form aggregates of various sizes, measured on a larger scale. Finally, depending on the nature of the interactions, these aggregates may associate to form agglomerates due to weak, often fragile, bonds. [17]

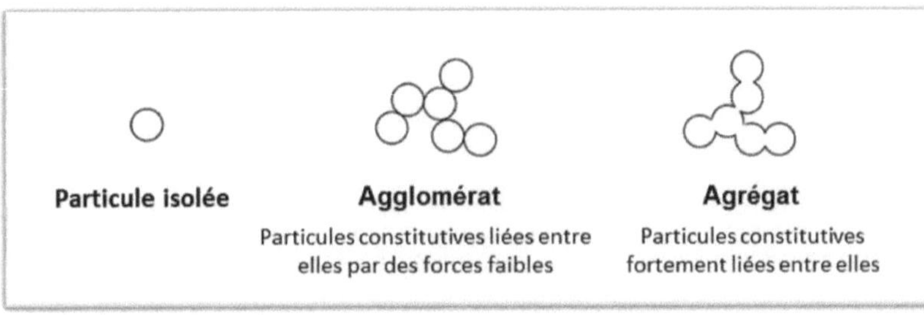

Figure I.2: Different load structures [17].

I.2.3 Load selection

The choice of filler depends on the properties required in the finished product. Generally speaking, fillers must meet the following requirements:

- Compatibility with base resins: non-toxicity, chemical inertness, absence of undesirable coloration, thermal stability, low water absorption and absence of influence on polymer structure and stability;
- Good wetting properties: optimal adhesion of filler to resin;
- Uniformity of quality and grain size;
- Low abrasive action on processing tools;
- Low cost: except for specific applications, such as aeronautics. [18]

I.2.4 Load classification

Expenses can be classified according to their nature and form:

I.2.4.1 Organic loads

A distinction is made between: [19]

- Natural vegetable fillers such as wood flour, cellulose fillers or cotton flour.
- Synthetic fillers such as silicas and nylon fibers.

I.2.4.2 Mineral fillers

Mineral fillers are frequently used in powder form: chalk (widely used), talc, carbon black, synthetic silica (thixotropic agent), mica, kaolin, barium sulfate, barium ferrite, calcium carbonate (widely used).

I.2.4.3 Reinforcing charges

The introduction of reinforcing fillers, such as carbon black or synthetic silicas, generally serves to improve modulus of elasticity, fracture resistance, impact strength, hardness and abrasion resistance. They can be used to modify electrical properties or also reduce permeability to gases and liquids. [19]

A distinction is made between: [20]

- **Fibrous reinforcing fillers**: mainly intended to improve mechanical properties;
- **Non-fibrous reinforcing fillers**: made up of very fine particles, such as non-hollow glass microbeads.

I.2.4.4 New expenses (2008)

Nanometric clay for nanocomposites and bottle packaging.

* Cereals, up to 50% incorporation, to improve recyclability.

* Lin, to increase the hardness of polypropylene. [16]

I.2.4.5 Load properties

- High hardness ;
- Chemical inertia ;
- Refractive index close to that of resin matrices ;

- Opacity controlled by addition of titanium dioxide pigments (TiO_2).

Increasing the filler percentage improves mechanical properties (especially if the filler content is > 60% by volume), and reduces polymerization shrinkage, coefficient of thermal expansion, absorption coefficient and water solubility.

Increasing the percentage of fillers and decreasing their size has the effect of improving the surface finish, which improves aesthetics and reduces the material's aggressiveness towards the periodontium, as well as increasing the material's wear resistance. [15]

I.3 Calcium carbonate

It's an inorganic compound naturally present in limestone, chalk and marble. It occurs as white powder or colorless crystals, virtually insoluble in water. This mineral is abundant on Earth, making up around 4% of the earth's crust. The main sources of calcium carbonate are deposits of the remains of marine organisms that have settled to the seabed. These organisms absorb calcium carbonate to form their skeletons and shells, contributing to the formation of rocks such as chalk, limestone and marble. [21]

It is an ionic compound that exists in six different allotropic forms: calcite, aragonite, vaterite, $CaCO_3$ monohydrate, $CaCO_3$ hexahydrate and $CaCO_3$ amorphous. [22]

The different natural forms of calcium carbonate, although chemically identical, vary in purity, whiteness and other characteristics. This compound is widely used in a variety of fields, including as a building material, in the food industry to deacidify wine and as a food additive. In addition, calcium carbonate is used in the beauty industry for the manufacture of cosmetics and toothpaste, as well as in agriculture as a fertilizer. [23]

Figure I.3: Calcium carbonate powder [22].

I.3.1 Polymorphism

Precipitated calcium carbonate exists in three forms: calcite, aragonite and vaterite. These polymorphs have the same chemical composition but different crystalline structures. Calcite is the most stable polymorph and is found in sedimentary rocks, while aragonite is synthesized at high temperature and pressure and is found in mollusc shells and metamorphic rocks. Vaterite is the rarest polymorph because it is unstable under normal conditions. [24]

Figure I.4: Microphotograph of $CaCO_3$ [25]

Calcium carbonate synthesis techniques include both physical and chemical methods. Physical techniques, such as mechanical grinding, are not effective due to the difficulty of obtaining particles of nanometric size and regular morphology. Chemical techniques, such as precipitation and emulsions, are advocated because of their ease of execution and low energy cost. [26]

Precipitation methods generally involve the preparation of a calcium hydroxide solution followed by its carbonation with carbon dioxide bubbles to form $CaCO_3$.

Another method involves mixing sodium carbonate with calcium chloride or calcium nitrate to also form $CaCO_3$. [27]

Emulsion methods involve mixing the organic phase with an aqueous solution of carbonate ions and another aqueous solution of calcium ions. Single emulsions involve mixing the two solutions and stirring to produce calcium carbonate, while double emulsions involve mixing the two emulsions and stirring to also produce calcium carbonate. These methods enable nanometric particles and various polymorphs of calcium carbonate to be obtained. [26-28]

Figure I.5: Natural rocks for calcium carbonate mining [29].

Table I.2: Different properties of CaCO polymorphs$_3$ [30]

Polymorph	Group	Density (g/cm)^3	Crystalline habitus
Calcite	R3c	2.71	104
Aragonite	Pmcn	2.93	110, 011, 010
Vaterite	/	2.66	/

I.3.2 Fields of application

I.3.2.1 Glassware

Calcium carbonate is an indispensable constituent in the manufacture of hollow glass, fiberglass, optical glass, glass wool and more. It gives glass improved heat resistance, better mechanical properties and greater resistance to chemical attack. [31]

I.3.2.2 Agriculture

Calcium carbonates (limestone, chalk, marl) and magnesium carbonates (dolomite) are widely used as soil improvers to correct soil acidity, improve soil texture and provide the calcium oxide (CaO) and magnesium oxide (MgO) nutrients required by plants. [31]

I.3.2.3 Animal and human nutrition

Carbonate powders are mixed with animal feed (cattle, pigs, poultry) for their calcium content.

In the human food sector, they are used in confectionery and bakery products as anti-caking agents and in the manufacture of chewing gum. [31]

I.3.2.4 Paint, rubber and plastics industry

In paints, calcium carbonate is used as a mineral filler to extend resins, aid pigmentation and control the finish.

In plastics, it is used as a filler up to 25% by volume, often mixed with additives.

In rubber, it helps improve mechanical properties and is used as a diluent. [31]

I.3.2.5 Paper

Calcium carbonate is used as a filler in paper, where it represents 15 to 40% of the composition. It reduces acidity and improves durability. [31]

I.3.2.6 Chemical industry

It is used as a raw material in the manufacture of other chemicals, such as soda ash and calcium chloride. It is also used for industrial gas desulfurization, water treatment and neutralization. [31]

I.3.2.7 Construction / Public works

As fillers (powders $<80\mu m$), carbonates are used in asphalt, road mixes, concrete, etc. They improve mechanical properties and cohesion. They improve mechanical properties and cohesion. [32]

I.3.2.8 Pharmaceuticals and cosmetics

In the form of very fine, pure powders (>98% $CaCO_3$), they are used as excipients in medicines, toothpastes, cosmetics and other products.

Users, especially in the public sector, generally require a calcium carbonate with at least 95% $CaCO_3$, with a whiteness of 80 to 98% depending on use, a variable moisture content and a suitable particle size. [31]

I.4 Composites

Composites are materials made up of two or more distinct phases, whose combined properties are better than those of the individual components taken separately. These materials are designed to achieve specific properties and optimized performance for a variety of industrial applications. Composites are used in many fields, including aerospace, automotive, construction and sports equipment. [33]

I.4.1 Types of composites

I.4.1.1 Metal matrix composites (MMC)

CMMs use a metal matrix reinforced with fibers or particles to improve mechanical properties and thermal resistance. [34,35]

I.4.1.2 Ceramic matrix composites (CMC)

CMCs use a ceramic matrix, reinforced to improve resistance to thermal and mechanical shock. [34,35]

I.4.1.3 Polymer matrix composites (PMC)

CMPs use a polymer matrix, reinforced with fibers or fillers to improve mechanical, thermal and other properties. [34,35]

I.4.1.4 CMP applications

Polymer matrix composites are the most commonly used because of their light weight, relatively low cost and ease of processing. They generally consist of a polymer matrix reinforced with fibers (such as glass, carbon or natural fibers) or fillers (such as mineral particles). [36]

a. Advantages of CMP

1. Improved properties: Increased stiffness, tensile strength and thermal stability. [37]

2. Cost reduction: Mineral fillers can reduce the overall cost of materials. [37]

3. Customization: Adjustment of material properties to meet specific requirements. [37]

4. Ecology: Using natural fibers and mineral fillers can improve durability and reduce carbon footprint. [37]

b. Types of loads used for CMP

Polymer matrix composites use various types of reinforcement to enhance their properties. These include mineral fillers such as calcium carbonate, talc, kaolin and silica. Fibrous reinforcements include glass and carbon fibers, as well as natural fibers such as hemp, jute and sisal. Nanoparticles such as nanoclays and carbon nanotubes are also used. Finally, functional additives such as UV stabilizers, antioxidants, flame retardants and plasticizers are added to optimize the performance of these composites. [38]

c. Techniques for preparing mineral-filled CMPs

❖ **Melt blending or direct blending**

Mineral fillers are mixed directly with the polymer melt in an extruder or internal mixer prior to the molding or extrusion process. This simple and efficient method ensures good filler dispersion. [39]

❖ **Masterbatch**

A masterbatch is a concentrate of mineral fillers dispersed in a small quantity of polymer. This concentrate is then mixed with the main polymer during the molding or extrusion process. This technique ensures homogeneous dispersion of the mineral fillers, facilitates precise dosage and improves the quality of the final composite. [40]

❖ **Mixing in solution**

Mineral fillers are dispersed in a polymer solution. The solvent is then evaporated to form the composite. This technique is suitable for soluble polymers that are difficult to blend in the molten state. [39,41]

- **Solid blending**
  - Dry mixing of powders before melting;
  - Simple, economical method.
- **Emulsion mixing**
  - Dispersion in a liquid phase, followed by coagulation and drying;
  - Useful for fine loads.

I.4.2 Masterbatches preparation techniques

Masterbatches can be prepared by : [42]

1. **Extruder**: The most commonly used technique is single-screw or twin-screw extrusion, where fillers and polymer are mixed at high temperature and under shear in an extruder. This method is particularly suited to large-scale production.
2. **Internal mixer**: This is the second option for preparing Masterbatches, where mixing takes place in a closed chamber under high shear forces. This technique is used for mixtures requiring a fine, homogeneous dispersion. It is suitable for viscous polymers and complex formulations.
3. **High-speed mixer**: performs rapid pre-dispersion prior to melting, offering a fast and economical method.
4. **Dry mixing**: Simple dry mixing of powders. It is used for premixes prior to heat treatment.

References

[1] : AquaPortail. "Polymer." [online]. Available from: <www. aquaportail.com> (Accessed: 16/05/2024).

[2] : D. DOLATKHANI. "What is a polymer?" [online]. Available at: <https://polymerexpert.fr>. (Accessed 05/16/2024).

[3] : StudySmarter. "Polymers: definition, types & properties." [online]. Available from: <https://www.studysmarter.fr> (Accessed: 16/05/2024).

[4] : Matechplas. "Polyethylene (HDPE)." [online]. Available at: <www.matieres techniques-plastiques.com>. (Accessed on: 20/05/2024).

[5] : Alamy. "High density polyethylene (HDPE)." [online]. Available at: <www.alamy.fr>. (Accessed: 05/02/2024).

[6] : C. PENU. "High-density polyethylene PE-HD: Manufacturing." [online]. Techniques de l'Ingénieur. Available at: <www.techniques-ingenieur.fr>. (Accessed: 20/05/2024).

[7] : R. DOSSOGNE. "HDPE. A 3315. Techniques de l'Ingénieur Paris, 1995, pp. 2-10.

[8] : J. A. BRYDSON, "Plastics materials. 7ème Edition, Ed. Butterworth-Heinemann, London, 1999.

[9] : T. SENNINGER. "Polymerization catalysis." [online]. Techniques de l'Ingénieur. Available at: <www.techniques-ingenieur.fr>. (Accessed: 05/21/2024).

[10] : Statista. "HDPE global production capacity by country 2021." [online]. Available at: <www.statista.com>. (Accessed: 04/07/2024).

[11] : A. VIVET. "Reinforcement of polymer matrices and composite materials by Nanoparticles", PhD Thesis. University of CAEN Basse, Normandie, 2012, 136 p.

[12] : R. DOSSOGNE. "High-density polyethylene PE-HD." [online]. Available at: < https://fr.scribd.com>. (Accessed: 05/21/2024).

[13] : J. AUCHER. "Comparative study of the behavior of thermoplastic and thermoset matrix composites." Mécanique, Paris: Ecole Doctorale SPMII, 2009.

[14] : E. TOSSOU. "Développement de nouveaux composites hybrides renforcés par des fibres de carbone et de lin: mise en œuvre et caractérisation mécanique", PhD thesis, University of CAEN, Normandy, Paris, 2019.

[15] : F. MESSAI. "Effect of surface treatment agent and accounting rate on the properties of thermoplastic/calcium carbonate matrix composites." Magister thesis, Université Ferhat ABBAS, Sétif, 2015. 228 p.

[16] : Materials. "Load. [in linen]. Available at: <http://fr.wikipedia.org> (Accessed: 30/03/2020).

[17] : N. JOUAULT. "Silica/polymer nanocomposites: filler structure, mechanical reinforcement, chain conformation and evolution under deformation." PhD thesis, Université de Bretagne-Sud, Paris, France, 2009, 277 p.

[18] : C. NAUDIN and C. CLOZZA. "Loads." [in French]. Available at: <www.techniques- ingenieur.fr> (Accessed: 22/04/2022).

[19] : F. ZOUKRAMI. "Rheological behavior and thermophysical, mechanical and microscopic properties of filled polymers in relation to interface and particle size." PhD thesis, Université Ferhat ABBAS, Sétif, Algeria, 2009, 197 P.

[20] : L. ZIADI. "Chemical modification of wood flour by adding a coupling agent: maleic anhydride-modified polypropylene PP-g-MA." PhD thesis, Abou al Kacem Saadallah University, Algiers, Algeria, 2004, 224 p.

[21] : Elémentarium, "Calcium carbonate. [online]. Available from: <https://lelementarium.fr>. (Accessed: 11/05/2024).

[22] : B.D. EIRASA and L. A. PESSANB. "Mechanical properties of polypropylene/calcium carbonate nanocomposites." Materials Research, Vol.12, 2009. pp. 517-522.

[23] : PCC Group. "What is calcium carbonate and what is it used for?" [online]. Available at: <www.products.pcc.eu/fr>. (Accessed: 11/05/2024).

[24] : B. ROMULAD. "Calcium carbonate nanoparticles." PhD thesis, Université de Laval, Québec, Canada, 2010, pp. 189.

[25] : Alamy. "Aragonite calcium carbonate $CaCO_3$." [online]. Available at: <www.alamyimages.fr>. (Accessed: 04/06/2024).

[26] : L. XIANG et al. "Influence of chemical additives on the formation of super-fine calcium carbonate." Powder Technology, Vol. 126, 2002, pp. 129-133.

[27] : S. H. WEI et al. "High surface area calcium carbonate: Pore structural properties and sulfation characteristics." Engineering Chemistry Research, Vol. 36, 1997, pp. 2141-2148.

[28] : T. HIRAI et al. "Biomimetic synthesis of calcium carbonate particles in a pseudo vesicular double emulsion. Biomimetic synthesis of calcium carbonate particles in a pseudo vesicular double emulsion." Langmuir, Vol. 13, 1997, pp. 6650-6653.

[29] : France Minéraux. "Characteristics of marble stone. [online]. Available at: <www.france-mineraux.fr>. (Accessed: 04/06/2024).

[30] : T. KATO, A. SUGAWARA, and N. HOSODA. "Calcium carbonate-organic hybrid materials." Advanced Materials, Vol. 14, pp. 869-877.

[31] : A. BELGHOUL and A. BOUACHA. "Contribution à la valorisation des carbonates de calcium", Mémoire d'Ingénieur, Ecole nationale polytechnique, Constantine, Algeria, 2018, 269 p.

[32] : N. POKORNIK. "Calcium carbonate in the concrete industry." [online]. Available at: <https://noahchemicals.com>. (Accessed: 04/06/2024).

[33] : W. D. CALLISTER and D. G. RETHWISCH. "Composites: properties, benefits and industrial applications." Materials Science and Engineering, Vol. 12, 2014, pp. 264-272.

[34] : B. HACHEMANE "Experimental analysis of the mechanical behavior of a composite material subjected to impact loading." PhD thesis, M'hamed Bougara University, Boumerdes, Algeria, 2013, pp. 146.

[35] : L. GENDRE. (2011), "Les grandes familles des matériaux composites." Paris, Edition Cachan.

[36] : M. J. JOHN and S. THOMAS. "Polymer matrix composites: lightness, cost and ease of processing." Carbohydrate Polymers, Vol. 71, 2008, pp. 343-364.

[37] : K. JAYARAMAN and D. BHATTACHARYYA. "Mechanical performance of woodfibre-waste plastic composite materials." Applied Composite Materials, Vol. 11, 2004, pp. 343-364.

[38] : K. VARAPRASAD et al. "Synthesis and characterization of hydrogel-silver nanoparticle-curcumin composites for wound dressing and antibacterial application." Journal of Applied Polymer Science, Vol. 115, 2010, pp. 1199-1207.

[39] : P. KAUSHIK and T. SABU. "Polymer-matrix nanocomposites, processing, manufacturing, and application: an overview." Advanced Materials Reviews, vol. 1, 2019, pp. 1900-1915.

[40] : L. MASCIA et al. "Masterbatch technology for filler addition in polymer processing: a review." Polymers, vol. 9, 2017, pp. 269-276.

[41] : T. SABU, J. KURUVILLA and J. S. K. MALHOTRA. "Polymer Composites." Macro- and Microcomposites, Vol.1, 2012, pp. 283-291.

[42] : K. R. PATEL et al. "Masterbatch Preparation: An Overview. Polymer-Plastics Technology and Materials." Vol. 57, 2018, pp. 535-555.

Chapter II: Experimental methodology

II.1 Materials used

II.1.1. High-density polyethylene

SABIC® HDPE F00952 stands out as an advanced HDPE copolymer designed specifically for blown film extrusion. Its unique combination of features makes it ideally suited to demanding applications: [1]

- Excellent extrudability due to high molecular weight and broad molecular distribution, guaranteeing easy processing into blown films;
- High-strength, high-density films ideal for durable shopping bags, heavy-duty packaging films and multi-wall bag liners;
- Presence of antioxidants for longer life of finished products, delaying degradation and preserving film integrity;

Beyond its outstanding properties, SABIC® HDPE F00952 offers additional advantages: [1].

- Easy film processing and printing for crisp, attractive graphics;
- Conversion to HMW-HDPE (High Molecular Weight High-Density Polyethylene standard) equipment without major modifications, simplifying integration into existing processes;
- Adapted melting temperatures (between 200 and 235°C) for efficient, energy-saving production;
- Excellent film properties (measured on 15 μm blown film) guarantee optimum performance in terms of strength, rigidity and durability.

Table II.1: Main characteristics of HDPE **[1].**

Properties	Values	Units	Test methods
Polymer properties			
Fluidity index			
At 190°C and 2.16 Kg	0.07	g/10min	ISO 1133
At 190°C and 21.6Kg	8.5	g/10min	ISO 1133

Density	950	g/cm^3	ISO 1133
Film properties			
Shock resistance	32	kJ/m^2	ASTM D4272
Film tensile test			
MD yield strength	30	MPa	ISO 527-3
Stress at break TD	50	MPa	ISO 527-3
MD breakup stress	40	MPa	ISO 527-3
Tensile strength TD	400	%	ISO 527-3
MD tensile strength	300	%	ISO 527-3
Modulus of elasticity TD	620	MPa	ISO 527-3
Modulus of elasticity MD	620	MPa	ISO 527-3

II.1.2. Calcium carbonate

Omyalite 90T is a fine calcium carbonate powder treated with stearic acid, designed to enhance product performance in a multitude of applications. Its chemical composition gives it remarkable properties, such as whiteness, opacity and compatibility with various polymers. [2]

This product acts as an effective filler, increasing the rigidity, tensile strength and flexural strength of materials, while reducing production costs. Omyalite 90T is particularly useful in the manufacture of plastic films and bags, pipes and fittings, automotive parts, building materials, cosmetics and cardboard, paints and coatings, adhesives and sealants, and rubber products. [2]

Its antistatic, flame-retardant and reinforcing properties make it the additive of choice for optimizing the performance of finished products. Environmentally friendly and biodegradable, Omyalite 90T is part of a responsible, sustainable approach. [2]

Table II.2: Characteristics of calcium carbonate [2].

Aspect form	Powder
Color	Blanche

pH	8.0
Melting temperature	800° C
Relative density	2.93 g/cm^3 at 25°C
Hydro solubility	Insoluble

II.2 Working methodology

II.2.1. Apparent density measurement of $CaCO_3$

Tamped density, or bulk density after compaction, is the mass of a granular or pulverized material divided by its volume after compaction. [3] It is measured in the following steps:

- ✓ One gram of the sample is introduced into a graduated cylinder.
- ✓ The specimen is subjected to vertical tapping to compact the material, reducing voids between particles.
- ✓ Once tapping is complete, the volume of material tamped is read directly from the test tube.
- ✓ The tapped density is then calculated by dividing the mass (1 gram) by the tapped volume read.

Note: *The measured bulk density is 1.25 g/cm^3.*

II.2.2. Preparation of HDPE/$CaCO_3$ composites by the direct method and by the

CaCO concentrate method$_3$ (masterbatch)

- **Internal mixer (Plastograph)**

The plastograph, or torque rheometer, plays an important role in the study of polymer processing properties. Its expertise lies in precisely monitoring the melting of plastic mixtures, providing valuable data on their behavior during manufacturing processes. [4]

Figure II.1: Photograph of the Brabender plastograph

By measuring the torque applied, the plastograph delivers an informative characteristic curve (Fig. II.1). The melting time, maximum torque, minimum torque and stability level provide information on melting speed, solid polymer strength, viscosity and melt behavior, respectively. [4]

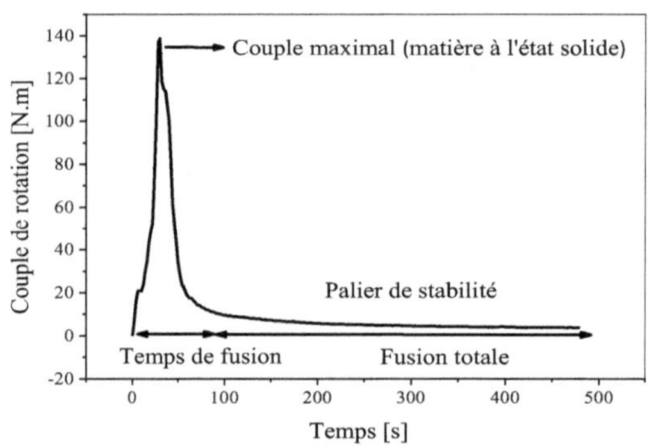

Figure II.2: General appearance of a plastogram

HDPE/$CaCO_3$ composites were prepared using two distinct approaches: the conventional method and the method using a charge concentrate (masterbatch).

In the conventional method, HDPE is first melted in the plastograph. Once the molten state has been reached, the $CaCO_3$ charge is added to the polymer matrix, and mixing is carried out in a closed chamber. In contrast, the masterbatch method relies on the prior preparation of a $CaCO_3$ concentrate, obtained by mixing a large amount of filler with a small amount of HDPE. This concentrate is then mixed with pure HDPE in appropriate quantities to achieve the desired $CaCO_3$ concentrations.

Both methods use the same operating conditions: a temperature of 220°C, a rotor speed of 40 rpm, a mixing time of 10 minutes, and a chamber volume of 55 cm^3. The batch was pre-steamed at 100°C for 24 hours.

II.2.2.1. Preparation of HDPE/CaCO mixtures₃ by the direct method

Four HDPE/$CaCO_3$ blend compositions were prepared based on the volume percentages of the components, calculated in relation to the total volume of the plastograph chamber (55 cm^3). In this approach, the volume is distributed so that, for example, for a mixture of 95% HDPE and 5% $CaCO_3$, 95% of the volume is occupied by HDPE and 5% by $CaCO_3$. The corresponding masses of each component were determined taking into account their respective densities, in order to completely fill the chamber volume.

Table II.3: Compositions of HDPE/CaCO blends$_3$

Percentage by volume	HDPE mass	CaCO mass $_3$
100 (%)	44 g	–
95 (%)	41.8 g	3.44 g
90 (%)	39.6 g	6.75 g
85 (%)	37.4 g	10.31 g
80 (%)	35.2 g	13.75 g

Note: *The exact mass of HDPE required to fill the 55 cm plastograph chamber (*^{3} *) is 44g.*

II.2.2.2. Preparation of HDPE/$CaCO_3$ mixtures by the concentrate method of

$CaCO_3$ (CC)

1. Preparation of $CaCO_3$ (CC) concentrate

- A mass (m_0) of HDPE is introduced into the plastograph.

- After melting, $CaCO_3$ is added gradually until saturation point is reached, i.e. when the HDPE can no longer absorb $CaCO_3$, leaving an unabsorbed excess.
- The excess $CaCO_3$ is removed using a soft brush, and the final product (HDPE/$CaCO_3$) is then weighed. This mass is noted (m_x).

- To determine the exact mass of $CaCO_3$ added, m_0 must be subtracted from m_x.
- Finally, the mass percentage of HDPE and $CaCO_3$ is calculated.

- The $CaCO_3$ concentrate prepared contains 30% HDPE and 70% $CaCO_3$.

2. Preparation of HDPE/$CaCO_3$ composites from concentrate (CC)

This table shows the quantities of $CaCO_3$ (CC) concentrate and pure high-density polyethylene (HDPE) required for the preparation of HDPE/CC composites, taking into account specific $CaCO_3$ concentrations. Calculations are made to ensure that the total mass obtained fills the plastograph chamber, respecting the respective densities of the two components. Composites are prepared on the same plastograph used to prepare HDPE/$CaCO_3$ composites by the direct method, under the same temperature and rotor speed conditions.

Table II.4: Quantities of CC and HDPE required for the preparation of HDPE/$CaCO_3$ composites at various $CaCO_3$ concentrations taking into account the densities of HDPE (0.8 g/cm³) and $CaCO_3$ (1.25 g/cm³).

Desired $CaCO_3$ concentration (%)	CC plug weight (g)	Mass of $CaCO_3$ contained in the CC (g)	Mass of HDPE contained in CC (g)	Mass of pure HDPE added (g)
5	4.92	3.44	1.48	41.8
10	9.83	6.88	2.95	39.6
15	14.75	10.31	4.44	37.4
20	19.64	13.75	5.89	35.2

Note: *For example, the first line of the table indicates that 4.92 g of concentrate, containing 3.44 g of $CaCO_3$ and 1.48 g of HDPE, should be used, then mixed with*

41.8 g of pure HDPE to obtain a final mixture composed of 95% HDPE and 5% $CaCO_3$, thus filling the total volume of 55 cm^3.

II.3 Characterization techniques

II.3.1. Measurement of the fluidity index

The melt flow index was measured to assess the flowability of the molten material. A Controlab Fluidimeter Model 5 was used to examine the impact of incorporating $CaCO_3$ on the rheological properties of HDPE. Tests were carried out in accordance with ISO 1133 (190° C, 21.16 kg), after drying all samples at 100° C for 24 hours.

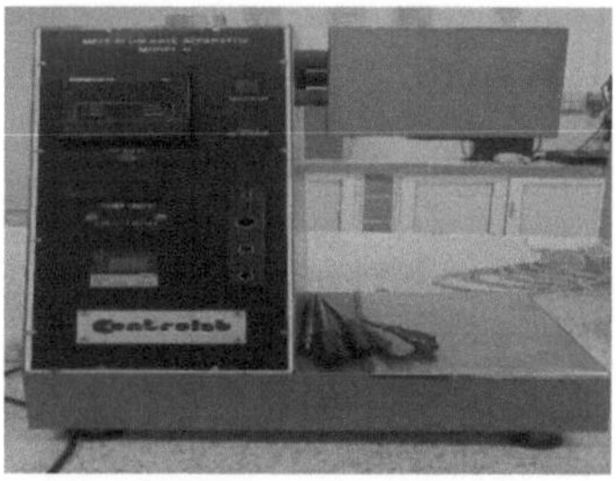

Figure II.3: Photograph of the Fluidimeter

The Fluidity Index is determined using the following formula:

$$\mathbf{IF\ (M, T) = 600\ \frac{m}{t}} \qquad \textbf{(Eq. 1)}$$

With : IF: Flow index (g/10min) ;

M : rated load (kg) ;

T: test temperature (°C) ;

600 is the reference time (s) ;

m : average mass of extrudates obtained (g) ;

t: time interval between two extrudate cuts (s).

II.3.2. Fourier Transform Infrared Spectroscopy (FTIR)

Infrared spectroscopy (FTIR) can be used to identify the functional groups present in the composite, thus verifying the chemical compatibility between high-density polyethylene (HDPE) and calcium carbonate ($CaCO_3$). The analyses were carried out using a QATR-S instrument. Films for FTIR testing were obtained by compression using a Polylab brand manual hydraulic press, at a temperature of 200°C, under a pressure of around 2 tons for 4 minutes. The process involved 2 minutes of preheating followed by 2 minutes of compression, with manual degassing carried out four times.

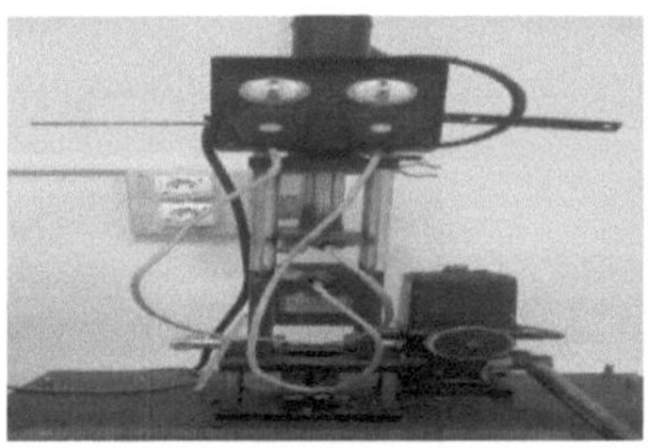

Figure II.4: Photograph of the hydraulic press

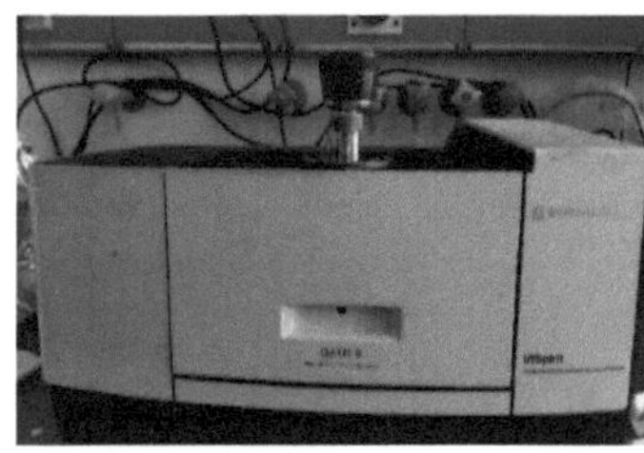

Figure II.5: FTIR photograph

II.3.3. Mechanical tests

Tensile and impact tests were carried out to assess the tensile and impact strength of the composites, providing vital information on their strength and durability. Specimens for these tests were prepared using a CARVER Model 30-12H hydraulic press, at a temperature of 200°C and a pressure of 200 kg/cm². This process lasted 12 minutes, comprising 7 minutes of preheating and 5 minutes of maximum compression, with two degassing stages. After 5 minutes of air cooling, the specimens were extracted by hand.

II.3.3.1. Tensile test

The tensile test is one of the most commonly performed tests, consisting of applying a uniform uniaxial tensile stress field to a specimen at a constant speed, while recording the resulting elongation until the material breaks. [5]

From the force-elongation curves, this test enables fundamental mechanical characteristics of the material to be determined, such as modulus of elasticity, stress at break (σr) and strain at break (εr). [6] Tensile tests were carried out at room temperature using a ZWICKRoell machine, model 445, equipped with a 10 kN force transducer, coupled with acquisition software. Tests were performed at a displacement speed of 20 mm/min. Specimens, 2.00 mm thick and 5.00 mm wide, were positioned vertically between two manually clamped jaws.

Figure II.6: Photograph of the traction machine

The force-elongation curves obtained made it possible to determine several fundamental mechanical characteristics, including yield strength, stress at break (σr) and elongation at break (εr).

a. Stress at break

The stress, which corresponds to the force 'F' applied to a section 'S', is calculated according to the following formula :

$$\sigma = \frac{F}{S}$$

(Eq. 2)

Where:

* σ: is the stress (MPa); F: is the force (N); S: is the cross-sectional area of the specimen (m^2).

b. Modulus of elasticity

Young's modulus 'E' (or modulus of elasticity) describes the relationship between stress and strain in the linear part of the stress-strain curve $\sigma = f(\varepsilon)$ It is calculated according to the following formula:

$$E = \frac{\sigma}{\varepsilon} \quad \textbf{(Eq. 3)}$$

Where:

* E is Young's modulus (MPa); ε is strain at break (%).

c. Deformation at break

Strain at break is determined by the following expression :

$$\varepsilon = \frac{\Delta L}{L} \quad \textbf{(Eq. 4)}$$

Where:

* ΔL: is the elongation, calculated as $\Delta L = L - L_0$ (mm); * L_0 : is the initial length of the specimen (mm); * L: is the final length of the specimen (mm).

II.3.3.2. Impact resistance test (Izod)

The impact test measures the total energy required to initiate and propagate a crack in a standard specimen under the effect of a calibrated impact force. The tests were carried out using a RESIL IMPACTOR machine. Specimens were V-notched, 0.5 mm deep, 2.20 mm thick and 12.50 mm wide.

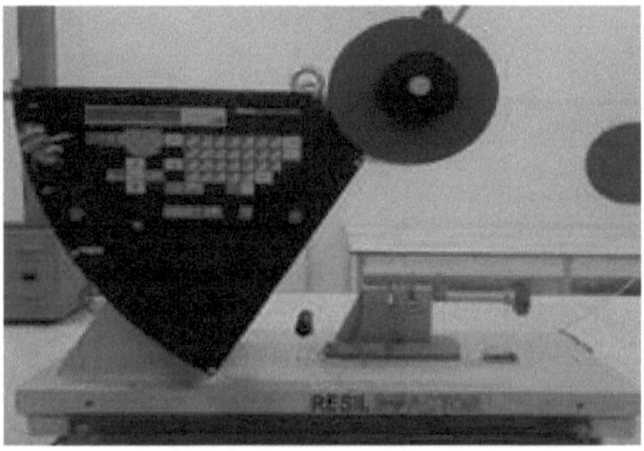

Figure II.7: Photograph of the shock machine

The impact strength Izod of notched specimens "a_k " is calculated according to the following equation:

$$\mathbf{a_k = \frac{Ec}{(L-0.5).\ e} x\ 10^{3}} \quad \textbf{(Eq. 5)}$$

With :

* a_k : impact resilience (kJ/m^2); * E_C : energy absorbed by specimen rupture in (J); * e : specimen thickness in (mm); * L : specimen width in (mm).

II.3.4. Determining the Vicat softening temperature

Vicat temperature, also known as Vicat softening point, designates the temperature at which a polymer begins to soften under a specific load when heated at a rate of 120°C/hr. This temperature is determined by observing the penetration of a circular probe into the material, subjected to a defined load, to a depth of 1 mm. Standardized by guidelines such as ISO 306 and ASTM D1525, this measurement is essential for assessing the thermal properties of polymers. It provides a crucial indication of the maximum temperature at which the material can be used under high-temperature point loading. [7,8] The test was carried out on an HDT/Vicat Ceast HV3 machine.

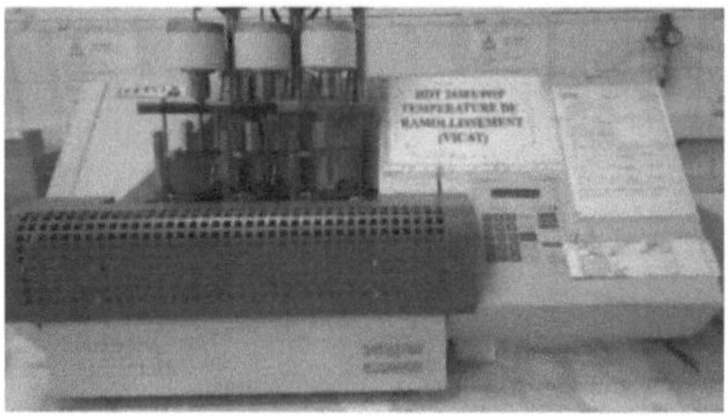

Figure II.8: Photograph of Vicat apparatus

The specimens used for this test were prepared under the same conditions as those used for the tensile and impact tests.

II.3.5. Thermal study: Differential Scanning Calorimetry (DSC)

DSC measures a material's thermal transitions, such as glass transition temperature (Tg), crystallization temperature (Tc), melting temperature (Tf), as well as the degree of crystallinity and signs of degradation (melting point drop). The technique is fast, economical and requires little material.

In this work, we limited ourselves to determining the melting temperature. A SETARAM DSC 131 EVO was used to characterize the samples, with a temperature range of 20°C to 450°C, a heating rate of 10°C/min, and a nitrogen atmosphere at a flow rate of 100 ml/min. Sample mass was 15 mg.

Figure II.9: Photograph of the DSC 131 EVO SETARAM

Experimental Protocol

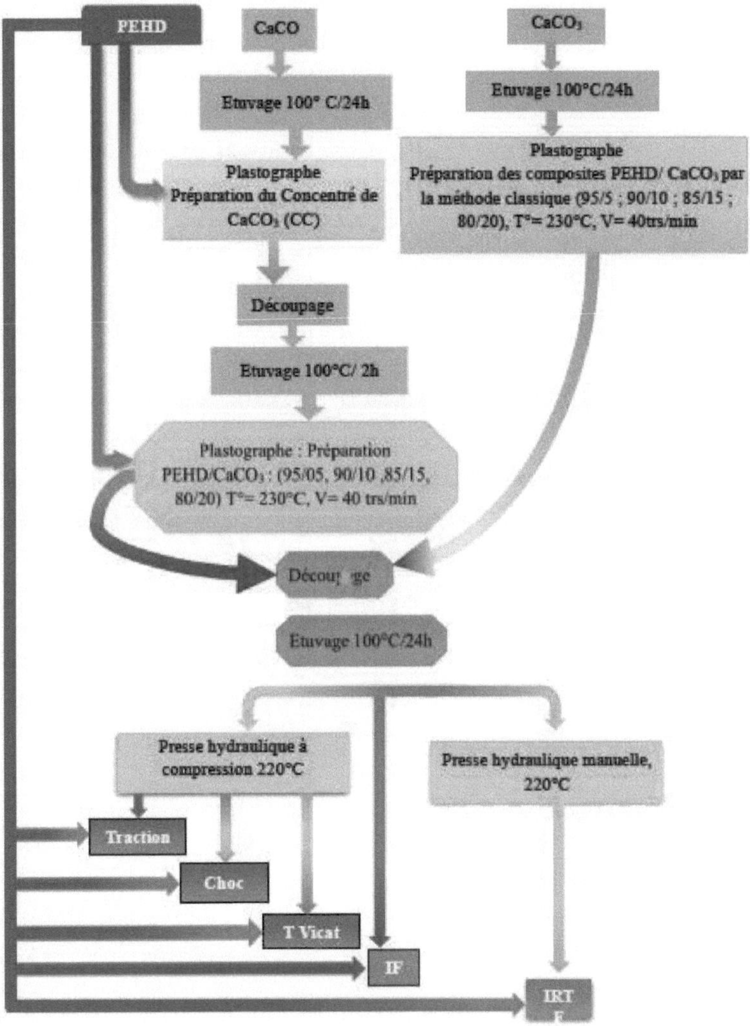

Figure II.11: Experimental protocol (Preparation of HDPE/CC mixtures)

References

[1] Data sheet supplied by Reverté SABIC® HDPE -F00950.
:

[2] Data sheet supplied by Reverté CARBONATOS DE CALCIOM.
:

[3] G. YORO and G. GODO. "Les méthodes de mesure de la densité apparente.
: Cahiers Orstom Série Pédologie, vol. 25, 1990, pp. 423-429.

[4] Thermoplast. "Quality and laboratory - TH-Extrusion." [online]. Available at:
: <https://th-extrusion.eu/fr>. (Accessed 05/16/2024).

[5] ISO 179-3. "Plastics - Determination of charpy impact characteristics. Part 2,
: instrumented impact test, 1993.

[6] F. ROUESSAC, A. ROUESSAC and D. CRUCHE. "Chemical Analysis. 6eme
: Ed., Dunod, Paris, 2004.

[7] Zwick Roell. "Deflection temperature under load HDT & Vicat softening
: temperature VST." [online]. Available at: <www.zwickroell.com>. (Accessed: 18/05/2024).

[8] Analytice. "Determination of the Vicat point in the laboratory. [online].
: Available at: <www.analytice.com>. (Accessed: 05/30/2024).

Chapter III: Results and discussion

III.1 Fluidity index results

- **HDPE/CaCO composites₃ prepared by both methods**

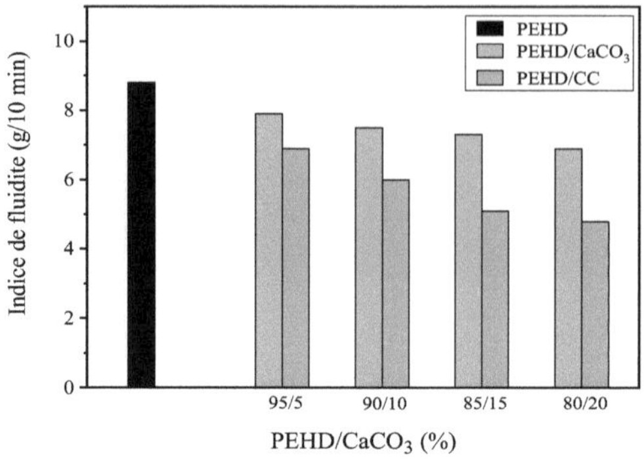

Figure III.1: Variation in melt flow index as a function of CaCO percentage $_3$

The reduction in flow index observed with increasing $CaCO_3$ concentration indicates that the latter acts as a filler, increasing the viscosity of the mixture and impeding the movement of polymer chains, a phenomenon typical in composites. The interaction between $CaCO_3$ and high-density polyethylene (HDPE) may also influence the structure of the polymer matrix, which could have consequences for the flow dynamics of the composite.

Furthermore, the difference in melt flow index values between the two methods can be interpreted as a consequence of the better dispersion of the filler within the matrix. Homogeneous dispersion is crucial, as it enables optimum interaction between $CaCO_3$ and HDPE, promoting uniform stress distribution within the composite. This can also reduce areas of stress concentration that might otherwise lead to defects during processing. As a result, improved dispersion can contribute to a greater increase in the composite's viscosity, thus impacting its mechanical and physical properties.

In addition, it's essential to consider that viscosity depends not only on filler concentration, but also on its morphology and size. Finer $CaCO_3$ particles can generate a larger contact surface with the polymer, improving dispersion and modifying the viscoelastic behavior of the composite. [1]

III.2 Results of Fourier transform infrared spectroscopy

III.2.1 HDPE characterization

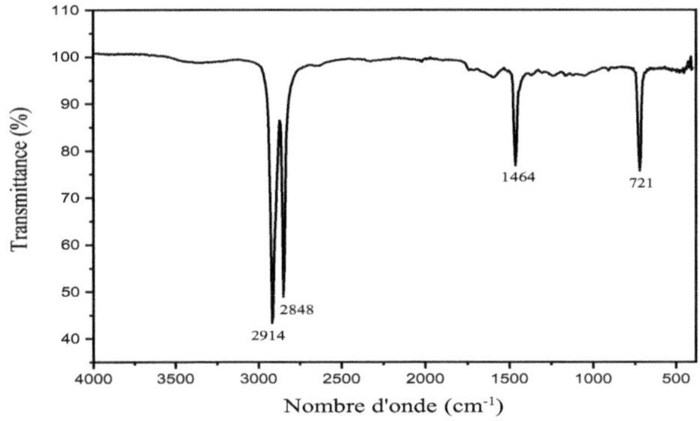

Figure III.2: Infrared spectrum of HDPE

The main bands characteristic of HDPE are summarized **in Table III.1**, and include two vibration bands for symmetrical and asymmetrical elongation of the C-H bonds of the -CH_2 - aliphatic group (2848- 2914cm^{-1}), two doublets which are merged with each other observed at 1472-1464 cm^{-1} and 730-721 cm^{-1} (Figure III.2.), corresponding to the crystalline part (1474 and 730 cm^{-1}) and the amorphous part (1464 and 720 cm). -1

Table III.1: Main vibrations of HDPE

Wave number $(cm)^{-1}$	Vibration type
2914	ν (as) C-H
2848	ν (s) C-H
1464	δ C-H (amorphous phase)
721	γ C-H (amorphous phase)

With :

- **ν :** elongation vibration ;
- **δ:** deformation vibration ;
- **γ :** balancing vibration.

III.2.2. Characterization of CaCO 3

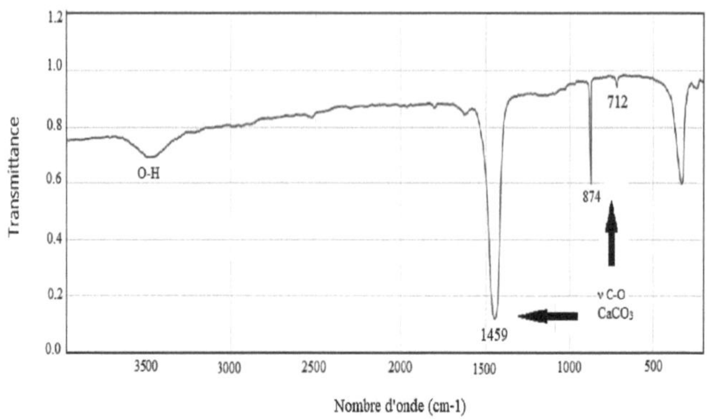

Figure III.3: Infrared spectrum of $CaCO_3$ [2]

Calcium carbonate is characterized by three modes of C-O elongation of the carbonate group, which appear as a triplet consisting of :

* A broad, intense absorption band at 1,459 or 1,462 cm^{-1} ; * A fine, intense band at 874 cm^{-1} ; * A fine, weak band at 712 cm $.^{-1}$

III.2.3. Characterization of HDPE/$CaCO_3$ and HDPE/CC composites

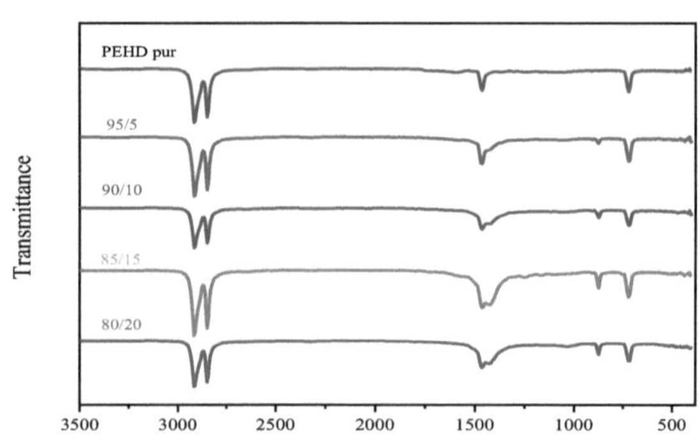

Figure III.4: Infrared spectra of HDPE/CaCO mixtures 3

Analysis of the infrared spectra of HDPE/$CaCO_3$ composites, prepared using both methods, revealed that the spectra are completely identical, with no notable differences. Thus, we decided to present only the spectra of composites obtained by method (CC). Figure III.4 shows these spectra in the range 450 cm-^{1} to 3500 cm-^{1}, which appear to include all the characteristic bands associated with the composites' constituent elements. In the rest of our work, these spectra will be examined in greater detail in different specific intervals.

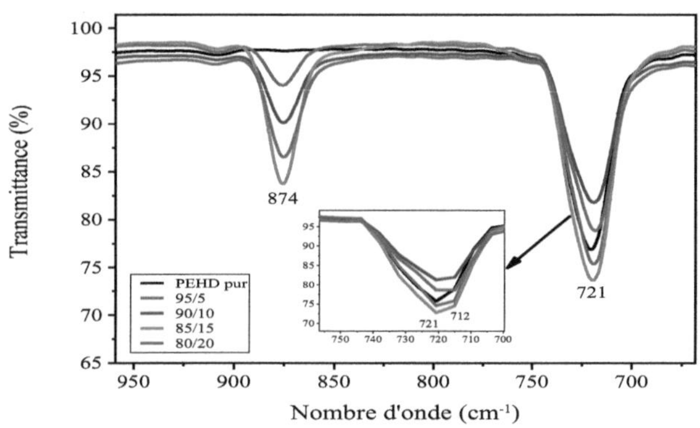

Figure III.5: Infrared spectrum of HDPE/CaCO composites $_3$ in the 600 cm range^{-1} - 950 cm^{-1}

As shown in Figure III.5, the infrared spectrum of HDPE/$CaCO_3$ composites in the 600 cm-^{1} to 900 cm-^{1} range reveals several key features. A distinct band at 721 cm-^{1}, corresponding to the rocking mode of the C-H bond in the HDPE structure, indicates preservation of the polymer matrix. A faint band at 712 cm-^{1}, associated with $CaCO_3$, appears only after zooming in on this specific region. The addition of $CaCO_3$ to the composite results in the appearance of a band at 874 cm-^{1}, related to the elongation of the carbonate C-O bond, whose intensity increases with the proportion of $CaCO_3$. This progressive intensification suggests good dispersion of the mineral filler and enhanced interaction with the matrix, recalling the observations of X. Zhang et al [3] in their FTIR study of layered silicate dispersion in polymer nanocomposites. In their case, for nanoclays in polymers, the increase in intensity and shifts of the FTIR bands indicated better delamination and integration. Similarly, the band evolution at 874 cm-[1] in HDPE/$CaCO_3$ composites can be interpreted as a sign of efficient and homogeneous $CaCO_3$ incorporation, providing indications of the quality of distribution and molecular interactions within the composite.

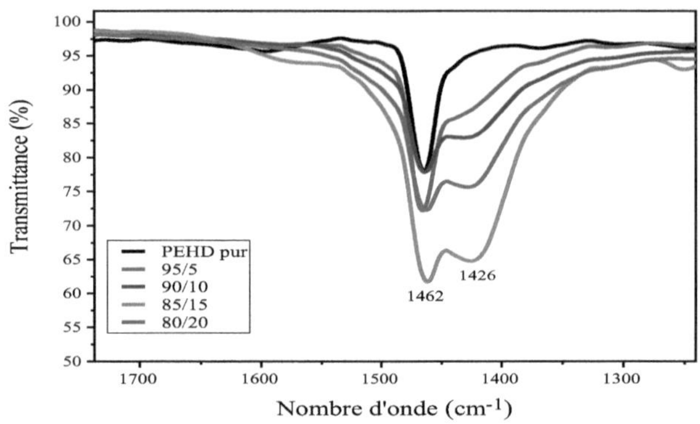

Figure III.6: Infrared spectra of HDPE/CaCO composites3 in the range 1250 cm^{-1} -1750 cm^{-1}

In this spectral region, a band is observed at 1462 cm-[1], corresponding to the vibration of the C-H bond of HDPE. In addition, a band located at 1426 cm-[1], associated with the C-O bond of carbonate, sees its intensity increase with increasing $CaCO_3$ content.

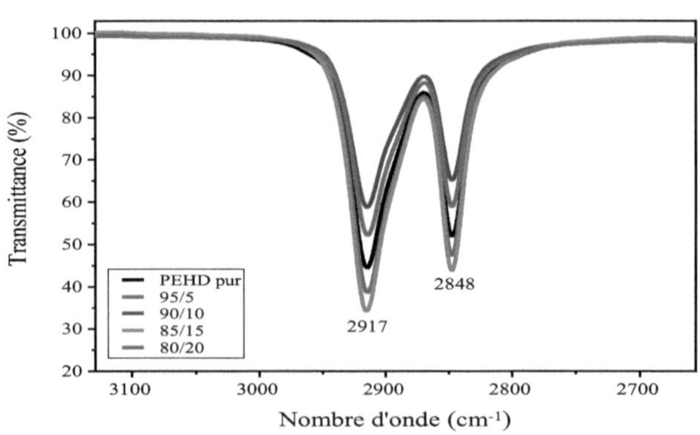

Figure III.7: Infrared spectra of HDPE/CaCO composites3 in the range 2650 cm^{-1} - 3150 cm^{-1}

In this spectral region, the characteristic bands of symmetrical and asymmetrical elongation vibrations of the C-H bonds of the aliphatic -CH_2-group are clearly visible and remain unchanged, suggesting that incorporation of $CaCO_3$ does not alter the overall shape of the infrared spectra. However, an overlap is observed between 1400cm-^{1} and 1470 cm-^{1}, where the bands of the C-H bonds overlap with those of the carbonate C-O bonds. This overlap could indicate good dispersion of $CaCO_3$ in the polymer matrix, in agreement with observations by X. Zhang et al. previously cited.

III.3 Tensile test results

III.3.1. Modulus of elasticity

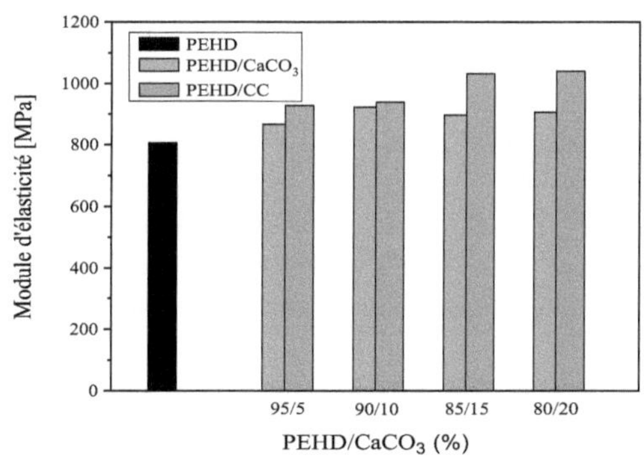

Figure III.8: Modulus of elasticity as a function of $CaCO_3$ percentage.

This curve reveals that the addition of $CaCO_3$ significantly increases the modulus of elasticity of HDPE/$CaCO_3$ composites, indicating an improvement in material stiffness. This is in line with typical behaviors of mineral-filler-reinforced polymers, where the polymer-filler interface contributes to limiting stress deformation. Increasing the $CaCO_3$ concentration from 5% to 20% leads to progressive stiffening, but a saturation effect is observed at high filler levels (15% - 20%), suggesting that the additional impact of filler on stiffness becomes marginal. This plateau may be linked to filler agglomeration beyond a certain threshold, which limits its effectiveness in the polymer matrix.

Comparatively, composites prepared from $CaCO_3$ concentrate (CC) show slightly higher moduli of elasticity than those prepared by the direct addition of $CaCO_3$. This difference can be explained by a potentially more homogeneous dispersion of $CaCO_3$ particles in the HDPE matrix thanks to the use of a concentrate. Indeed, better dispersion minimizes zones of weakness and optimizes stress transfer between filler and polymer. This observation highlights the importance of the preparation method in optimizing the mechanical properties of composites.

III.3.2. Maximum stress

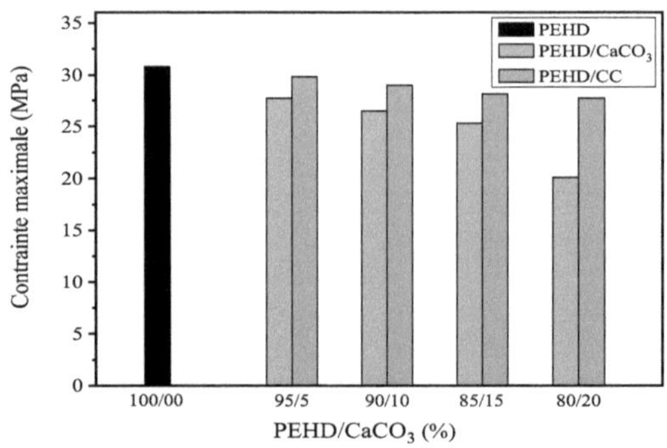

Figure III.9: Evolution of maximum stress as a function of $CaCO_3$ percentage.

The curve illustrates that incorporating $CaCO_3$ into HDPE/$CaCO_3$ composites results in a decrease in maximum stress, irrespective of the preparation method chosen. This indicates a reduction in strength under maximum load as the $CaCO_3$ concentration increases. Pure HDPE exhibits the highest maximum stress, reaching around 30 MPa. In contrast, the addition of $CaCO_3$ causes a progressive decrease in this stress, particularly marked at concentrations of 15% and 20%. This decrease could be due to the agglomeration of $CaCO_3$ particles, which introduces zones of weakness into the composite matrix.

III.3.3. Deformation at fracture

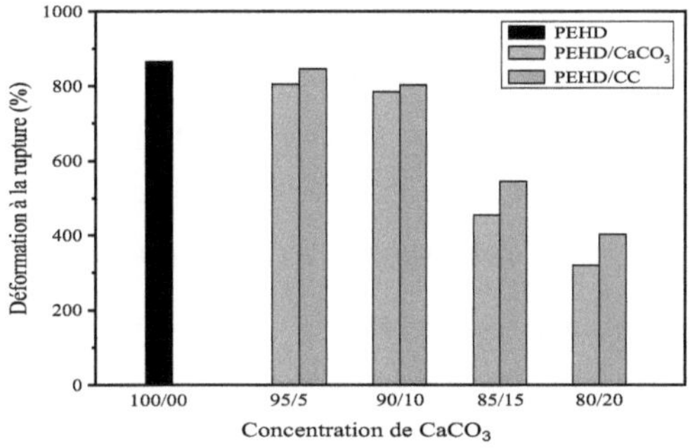

Figure III.10: Evolution of elongation at break as a function of $CaCO_3$ percentage.

The strain-at-break curve shows that the addition of $CaCO_3$ to HDPE progressively decreases its strain-before-break capacity, consistent with the increase in modulus of elasticity and decrease in maximum stress observed previously. Pure HDPE exhibits high strain at break, around 865%, indicating good ductility, while the addition of $CaCO_3$ makes the material stiffer and more brittle, reducing its ability to withstand significant stretching. At low concentrations (5% and 10%), the difference between the two preparation methods (direct and concentrated $CaCO_3$ addition) is minimal. However, at higher concentrations (15% and 20%), composites prepared with $CaCO_3$ concentrate show better deformation at break than those prepared by direct addition. This better behaviour could be attributed to a more homogeneous dispersion of filler particles in the case of the concentrate, limiting stress concentration zones and delaying fracture initiation. In sum, the use of $CaCO_3$ concentrate moderates the filler-induced loss of ductility, thereby optimizing the compromise between stiffness and fracture toughness for HDPE/$CaCO_3$ composites.

In conclusion, tensile testing of HDPE/$CaCO_3$ composites shows that the addition of $CaCO_3$ increases the modulus of elasticity, signifying a progressive stiffening of the material, but reduces the maximum stress and strain at break, indicating a loss of ductility and embrittlement of the composite. The use of a $CaCO_3$ concentrate facilitates the incorporation of the filler into the HDPE matrix, enabling the elastic modulus to be enhanced without significantly compromising mechanical properties at low filler concentrations. At higher concentrations, however, the embrittlement effect becomes unavoidable, as the accumulation of $CaCO_3$ introduces zones of weakness that reduce the composite's deformation capacity. These results highlight that while the addition of $CaCO_3$ improves the stiffness of HDPE composites, this is at the expense of strength and ductility, and that the concentrate incorporation method can better manage this balance within certain concentration limits.

III.4 Impact resistance test (Izod)

According to Figure III.11, impact strength remains almost unchanged for the 5% and 20% concentrations, whereas for the 10% and 15% concentrations the material becomes brittle.

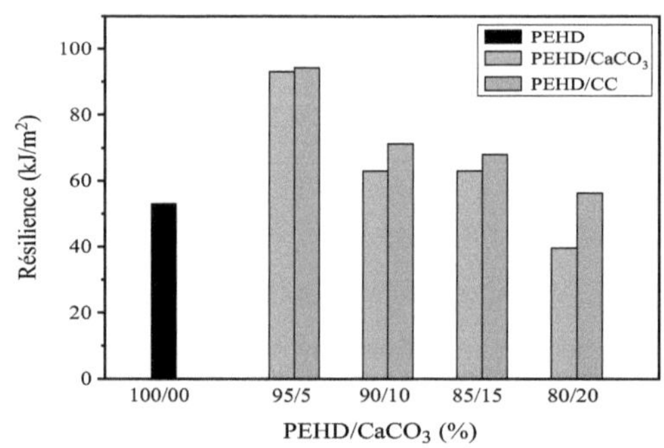

Figure III.11: Variation in impact strength as a function of $CaCO_3$.

Incorporation of $CaCO_3$ into the HDPE matrix by the concentrate method (CC) reaches a peak in impact strength at 5% load, with a value of 94.2 kJ/m^2, marking a significant improvement over pure HDPE (53 kJ/m^2). Although a gradual decrease in resistance is observed above 5%, values remain higher than those for pure HDPE. A similar behavior is observed for composites prepared by direct addition, where impact strength also reaches a maximum at 5%, then gradually decreases, remaining higher than that of pure HDPE up to 15%. Above this concentration, a marked drop to 39.52 kJ/m^2 is observed, which could indicate the formation of agglomerates. These results suggest that, although mechanical performance decreases beyond 5%, the composite retains satisfactory properties even at high $CaCO_3$ levels. Dispersion efficiency and the favorable interaction between polymer matrix and filler, particularly with the CC method, contribute to this improvement.

III.5. Vicat test results

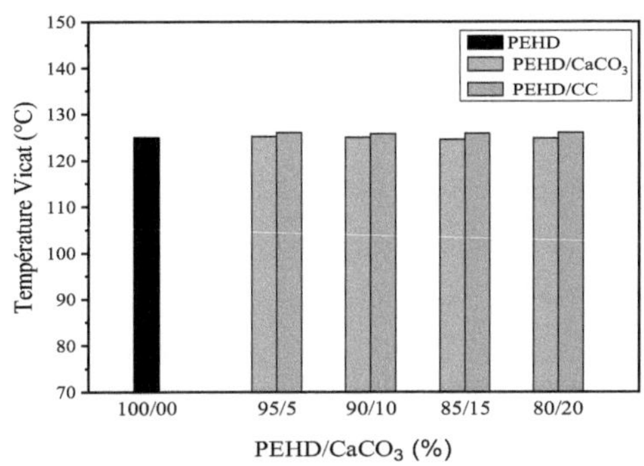

Figure III.12: Variation in Vicat temperature as a function of $CaCO_3$

The addition of $CaCO_3$ to HDPE/$CaCO_3$ composites causes a slight increase in Vicat temperature, ranging from 124.5°C to 126°C, even with $CaCO_3$ proportions of up to 20%. This rise is modest and relatively stable, whatever the preparation method. Both methods give similar results, with no significant differences. These observations suggest that the incorporation of $CaCO_3$, irrespective of the method chosen, does not substantially affect the thermal resistance of HDPE composites, thus enabling thermal stability close to that of pure HDPE (125°C) to be maintained.

III.6. DSC results

Table III.2: Influence of preparation method on melting temperatures HDPE/$CACO_3$ composites

Composition PEHD/$CaCO_3$ (%)	Melting temperature (°C) Method Concentrate (CC)	Melting temperature (°C) Direct method
Pure HDPE	145	145
95/5	145	145

90/10	147	145
85/15	148	142
80/20	148	146

DSC results show that the melting temperature of HDPE/$CaCO_3$ composites depends on the $CaCO_3$ incorporation method. For low concentrations (5%), the melting temperature remains constant at 145°C for both methods, indicating a negligible influence on the HDPE crystal structure. However, from 10% $CaCO_3$ upwards, a difference emerges: the concentrate method (CC) results in a gradual increase in melting temperature (up to 148°C at 20% $CaCO_3$), while the direct method shows less stability, with a slight decrease to 142°C at 15% $CaCO_3$. This difference suggests that incorporation by concentrate promotes a more homogeneous dispersion and better compatibility between $CaCO_3$ and HDPE, thereby enhancing the crystalline structure and thermal stability of the composite. In comparison, the direct method could result in less efficient dispersion, creating heterogeneities in the polymer matrix, affecting the melting temperature at high $CaCO_3$ concentrations.

References

[1] F. MOREL. "Understanding interfacial phenomena in composites based on
: precipitated calcium carbonate fillers: influence of surface treatment and processing method." PhD thesis, University of Lyon, France, 2010, 290 p.

[2] Nist. "Calcium carbonate." [online]. Available at: < https:// webbook.nis
: (Accessed: 06/30/2024).

[3] X. ZHANG, S. BHUVANA, L. S. LOO. "Characterization of Layered Silicate
: Dispersion in Polymer Nanocomposites Using Fourier Transform Infrared Spectroscopy." Journal of Applied Polymer Science, Vol.125, 2012, pp.175-180.

General conclusion

General conclusion

In conclusion, the experimental results obtained for HDPE/$CaCO_3$ composites prepared by two distinct methods (direct and concentrated $CaCO_3$ addition) revealed several significant trends concerning their mechanical, flowability and molecular interaction properties, as well as their thermal characteristics.

The addition of $CaCO_3$ led to an increase in the viscosity of the composite, as indicated by the reduction in the flow index with increasing mineral filler. This phenomenon, due to the reinforcing effect of the filler, shows how increasing $CaCO_3$ concentration results in greater flow resistance of the material, typical of filler-reinforced composites. Filler dispersion in the polymer matrix is also a key factor, as it influences flow dynamics and stress distribution, reducing stress concentration zones and improving overall mechanical properties.

Infrared spectra showed that the interaction between HDPE and $CaCO_3$ is effective, without significant modification of the polymeric structure. Intensity variations in characteristic bands, such as those of the C-O bond of $CaCO_3$, highlighted good filler dispersion and optimal interactions between carbonate and polymer, promoting homogeneous integration of the components and, consequently, a more robust composite.

Mechanical tests confirmed the impact of $CaCO_3$ on composite properties. The increase in modulus of elasticity, while beneficial in terms of stiffness, was accompanied by a decrease in maximum stress and strain at break. These trends suggest that, although strengthening is observed, the introduction of $CaCO_3$ also leads to material embrittlement, especially at high load concentrations. However, the use of $CaCO_3$ concentrate made it possible to limit this embrittlement effect, notably by ensuring better filler dispersion, thus optimizing the balance between stiffness and fracture toughness. This phenomenon is particularly notable for $CaCO_3$ concentrations below 15%, where incorporation by concentrate preserves good ductility and reduces the loss of mechanical performance.

Thermal results obtained by DSC revealed that the melting temperature of the composites depends on the method of $CaCO_3$ incorporation. For low concentrations of $CaCO_3$ (5%), the melting temperature remains stable at 145°C for both methods. However, from 10% $CaCO_3$ upwards, a difference became apparent: the concentrate method leads to a gradual increase in melting temperature, reaching 148°C at 20% $CaCO_3$, while the direct method shows less stability with a slight decrease in melting temperature to 142°C at 15% $CaCO_3$. This indicates that the concentrate method promotes better dispersion of $CaCO_3$, improving the crystalline structure and thermal stability of the composites.

As far as Vicat temperature is concerned, the addition of $CaCO_3$ caused a slight increase in this temperature, varying between 124.5°C and 126°C, regardless of the preparation method. This modest rise suggests that the incorporation of $CaCO_3$ does not significantly affect the thermal resistance of HDPE composites, thus enabling thermal stability close to that of pure HDPE to be maintained.

Finally, the results obtained in impact resistance tests completed this analysis, highlighting the way composites react under dynamic loading. They underline that the preparation method plays an essential role in optimizing mechanical properties, in particular the ability to retain good impact resistance despite stiffening of the composite.

Overall, these results suggest that the $CaCO_3$ concentrate method offers notable advantages, particularly with regard to filler dispersion, thermal stability and the balance between stiffness and mechanical performance. This makes it possible to design HDPE/$CaCO_3$ composites that are more homogeneous and better suited to applications requiring both optimized mechanical properties and good thermal resistance.

Outlook

Outlook

As part of the prospects for this work, it would be interesting to complement the analysis of HDPE/$CaCO_3$ composites with in-depth thermal studies, in particular by differential scanning calorimetry (DSC) and thermogravimetry (TGA). These techniques would enable us to better characterize the thermal properties, stability and degradation of composites, providing information on melting, transition and thermal degradation temperatures. In addition, a more detailed analysis of $CaCO_3$ dispersion in the polymer matrix could be carried out using scanning electron microscopy (SEM). This approach would offer a better understanding of the interaction between polymer and filler, while also enabling the examination of possible areas of weakness or stress concentration on fractured surfaces after mechanical testing.

In addition, simulations and modelling of the mechanical and thermal properties of HDPE/$CaCO_3$ composites could offer predictions of their behaviour under different conditions of use, while optimizing material formulation according to the requirements of industrial applications. Finally, it would be relevant to assess the environmental impact of composites, in particular by analyzing their recyclability and biodegradability. This would make it possible to determine whether these materials can meet the growing demands of sustainability and eco-design, offering more environmentally-friendly solutions for large-scale applications.

MIX
Papier aus verantwortungsvollen Quellen
Paper from responsible sources
FSC® C105338

Printed by Books on Demand GmbH, Norderstedt / Germany